Abdelghani Lakel
Sabrina Roguai

# GENERAL INFORMATION ON CARBON NANOTUBES

Abdelghani Lakel
Sabrina Roguai

# GENERAL INFORMATION ON CARBON NANOTUBES

ScienciaScripts

**Imprint**

Any brand names and product names mentioned in this book are subject to trademark, brand or patent protection and are trademarks or registered trademarks of their respective holders. The use of brand names, product names, common names, trade names, product descriptions etc. even without a particular marking in this work is in no way to be construed to mean that such names may be regarded as unrestricted in respect of trademark and brand protection legislation and could thus be used by anyone.

Cover image: www.ingimage.com

This book is a translation from the original published under ISBN 978-3-330-87193-9.

Publisher:
Sciencia Scripts
is a trademark of
Dodo Books Indian Ocean Ltd. and OmniScriptum S.R.L publishing group

120 High Road, East Finchley, London, N2 9ED, United Kingdom
Str. Armeneasca 28/1, office 1, Chisinau MD-2012, Republic of Moldova, Europe
Managing Directors: Ieva Konstantinova, Victoria Ursu
info@omniscriptum.com

Printed at: see last page
**ISBN: 978-620-8-39993-1**

# CONTENTS

# I - INTRODUCTION

Polymer composites have often successfully replaced metallic and ceramic materials. In many applications, polymer materials have offered an alternative to glass (PMMA, PC) and aluminium in aeronautics, which have been replaced by polymer matrix composites. Because of their lightness and ease of processing, polymers have long been used as electrical insulators, and the first work on extrinsically conductive polymers or polymers loaded with electrically conductive particles was carried out in the 1950s (Hitory of polymeric composites).

In composites with an insulating matrix loaded with conductive particles, the transport of electric current is frequently confined to metallic or graphite particles. The shape of the conductive phase is often granular (carbon black, silver platelets, etc.) or fibrous, such as carbon fibres, aluminium fibres, etc.

Structural composites are structures using high-performance composite materials that often replace metal structures, opening up new horizons for research into the use of this type of composite. However, these new horizons are synonymous with new problems, which more often than not lead designers to find solutions for integrating new functions into composite materials that were initially confined to mechanical strength.

## I - 1- Graphite

Graphite is a black solid with a hexagonal structure as shown in Figure 1. The planes, known as graphene planes, are stacked in an ABAB mode, with a distance between planes of 3.35 Å. These planes interact with each other only through attractive van der Waals forces. As these forces are relatively weak, the graphene planes can slide over each other. Within a plane, the length of the C-C bond is 1.415 Å.

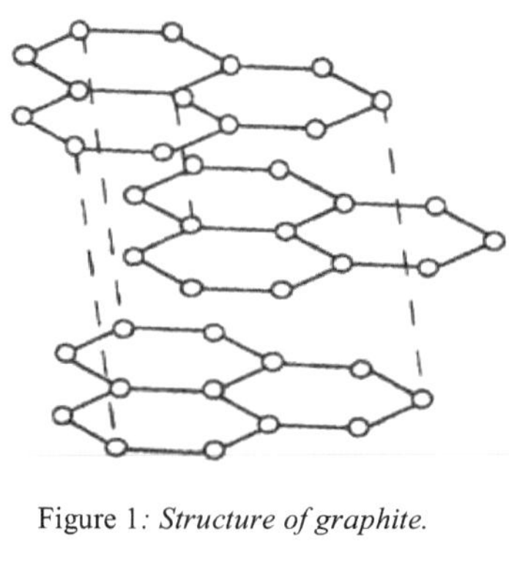

Figure 1*: Structure of graphite.*

Graphite can be available naturally or synthesised. Synthetic graphite is made from organic precursors. The precursor is first subjected to a carbonisation process, consisting of gradual heating in an inert atmosphere until temperatures of almost 1,300°C are reached. At this stage, amorphous carbon and polycrystalline graphite remain. A heat treatment at 2,500°C then enables a
graphitisation of this carbon to produce graphite.

Graphite is a good conductor. All the carbons are sp-hybridised[2], allowing delocalisation of π electrons and electrical conduction within a plane. Although there is no gap in its band structure (the conduction band overlaps the valence band. of 36 mev), [01] graphite does not conduct as well as metals because of its low charge carrier density. [02] It is therefore a non-metallic conductor, sometimes also called a semi-metal.

Graphite can also exist in rhombohedral symmetry (ABCABC stacking). This form of graphite is unstable compared with hexagonal graphite, which it can give after heat treatment at 1300°C. It never occurs alone, but coexists with the hexagonal form in both natural and synthetic graphite.

### I - 2- Diamonds

Diamond is the allotropic form containing only $sp^3$ carbons. It consists of atoms bonded as shown in Figure 2. The C-C bonds are 1.514 Å long. Diamond is thermodynamically unstable compared with graphite (the difference in free enthalpy is 2.88 $kJ.mol^{-1}$ ), but the diamond-to-graphite transformation is extremely slow under standard conditions. There are three main types of diamond: natural, synthetic and CVD.

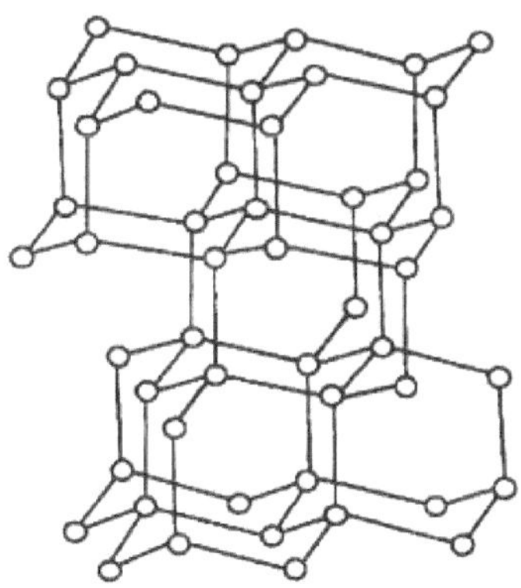

Figure 2: *Diamond structure.*

Like graphite, diamond also has another metastable crystalline form: a hexagonal form known as lonsdaleite. It should be noted that diamond is much denser than graphite (3.51 compared with 2.22).

Diamond is the hardest material and has found numerous applications as a high-performance abrasive. It has the highest known thermal conductivity of any solid at room temperature. It is an excellent semiconductor when its lattice is doped with small quantities of lithium, boron, nitrogen or phosphorus, and has a high refractive index, transmitting light from the far infrared to the ultraviolet.

### I - 3 - Fullerenes

Consider graphite, the thermodynamically stable phase of carbon: each of its graphene planes has a very large number, say an infinite number, of carbon atoms. If this number of atoms is greatly reduced, the number of dangling bonds on the edges of a graphene plane becomes high: the structure must then "close" in on itself. If the number of carbons is less than a few hundred, stable structures are closed: these are fullerenes, with the empirical formula $C_{2n}$. As a direct consequence of Euler's theorem, we can calculate that 12 pentagons are needed to give the curvature that allows the structure to close in on itself. [03]

With 12 pentagons and 20 hexagons, we form the famous $C_{60}$, or buckminster fullerene, named after Buckminster Fuller (the architect who designed geodesic domes, spatial structures offering a large volume for unequalled strength and lightness). This fullerene is shown in Figure 3 with its two types of C-C bonds: those common to two hexagons are called 6-6 bonds, those common to a hexagon and a pentagon are called 5-6 bonds. Due to the electron delocalisation of the π electrons, the carbons are all chemically equivalent: a $^{13}C$ NMR spectrum of $C_{60}$ in solution has only one peak.

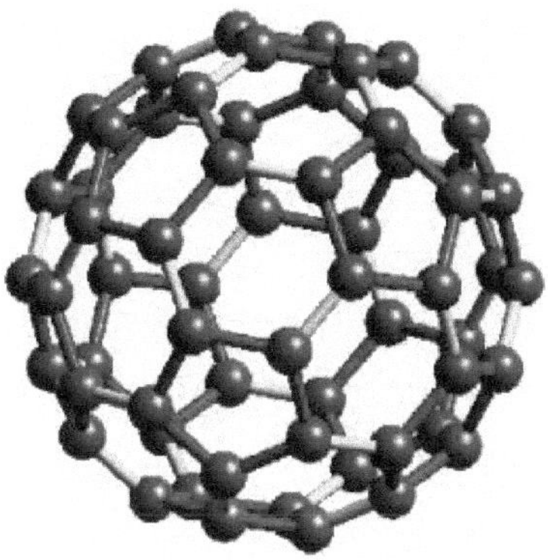

Figure 3: *Structure of $C_{60}$. The C-C bonds common to a hexagon and a pentagon are shown in red, the bonds common to two hexagons are shown in yellow.* [04]

## II - CARBON NANOTUBES

### II - 1 - Allotropic forms of carbon

Science thought it had come to the end of carbon's textural and structural possibilities, but that was to belittle Nature's inventiveness. After diamond and graphite (Figure 1), a new variety of pure carbon, fullerenes, whose compact stacking constitutes a new allotropic form of carbon, was synthesised in 1985 by Kroto et al [05]. This species takes the form of a closed structure of 60 carbons (C60) reminiscent of a football (figure 4).

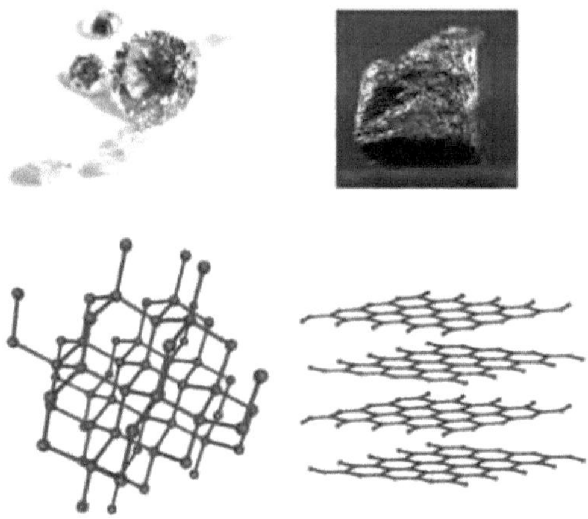

Figure 4: *Schematic representation of diamond (left) and graphite (right)* [06].

These tubes have a wall composed of certain layers of graphene in a concentric arrangement, spaced at a distance of 3.4 Å (Figure 5). The carbon atoms within a wall are sp-hybridized$^2$ . Tubular, multilayered "carbon filaments" had already been observed by TEM as early as 1958 [07], but the resolution of the apparatus did not allow the arrangement of the graphenes in relation to the elongation axis to be determined. With the stable increase in TEM determination [08].

The scientific community was shocked by the discovery of the first single-walled carbon nanotubes [09, 10], formed spontaneously during electric arc synthesis in 1993.

SWCNTs (single walled carbon nanotube) are made up of a single nanotube (figure 5), and spontaneously assemble in the form of bundles [11, 12]. The distance between two nanotubes, governed by Van der Waals interactions, is 3.15 Å [13]. Another form of nanostructured carbon, called carbon nanohorns (CNH), was also synthesised by Iijima in 1999 (Figure 6), and has properties similar to those of carbon nanotubes.

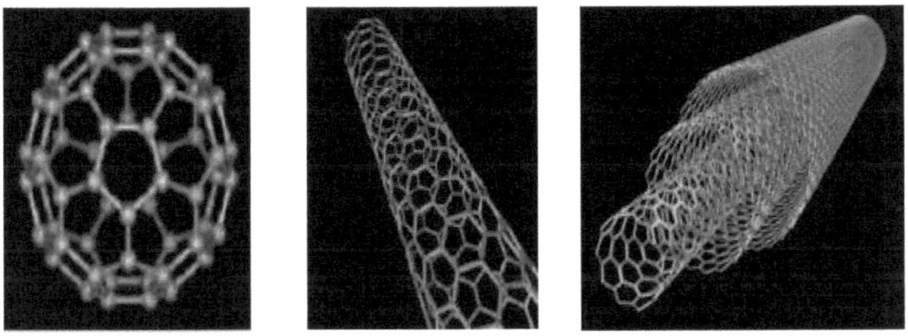

(a) (b) (c)

Figure 5: *Model structure of a* $C_{60}$ *(a), single-walled carbon nanotube (b) and multi-walled carbon nanotube (with three layers) (c)* [14].

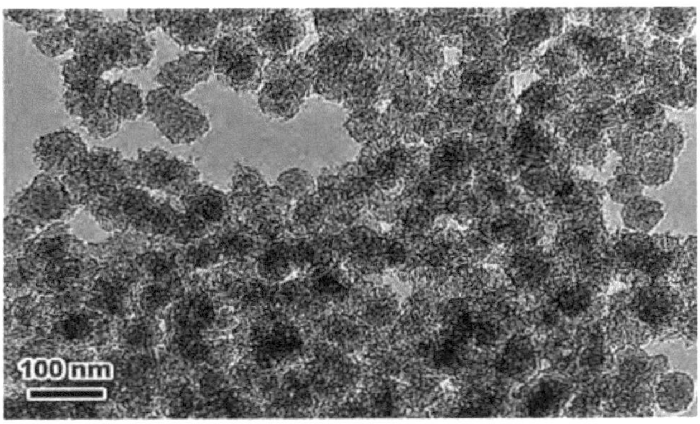


Figure 6: *TEM image of carbon nanocornet.*

Since then, numerous methods for synthesising these allotropic varieties of carbon have been developed. Interest in SWCNT and MWCNT (multiwall carbon nanotube) carbon nanotubes has grown frequently and a wide range of research on their properties, structure and various potential applications has been published.

## II - 2 - Geometric **structure**

Carbon nanotubes have a cylindrical shape and, depending on the case, the ends may or may not be closed by half-spheres. Their diameter varies between 1.4 nm and several nanometres and their length is of the order of 100 µm (this length can reach 1mm). The final structure of the carbon nanotube is formed by arrangements of graphene arranged in concentric layers, and depends on the direction in which the graphene is wound. The pair of integers (n, m), known as the Hamada index [45], are defined as the coordinates of the vector (Figure 7):

$$\vec{C}_h = n\,\vec{a}_1 + m\,\vec{a}_2$$

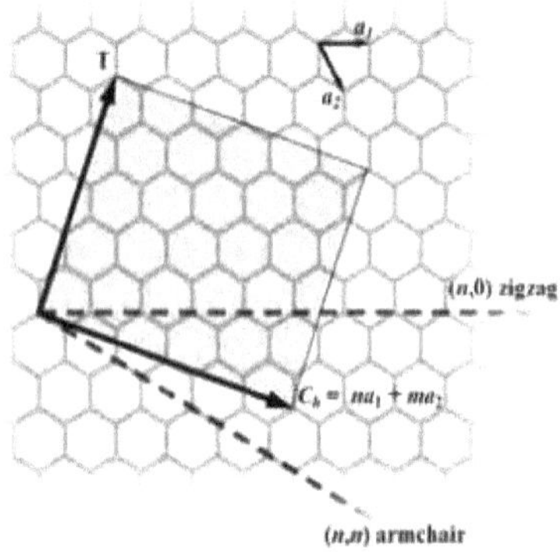


Figure 7: *Representation of the construction of a nanotube from a sheet of graphene* [15].

Depending on how the rectangle folds, the value of the pair (n, m) changes and the nanotube indicates a different geometry. In the case where n = m the geometry is called "armchair", in the case (n, 0) or (0, m) a zigzag geometry and in all other cases a chiral type geometry (figure 8).

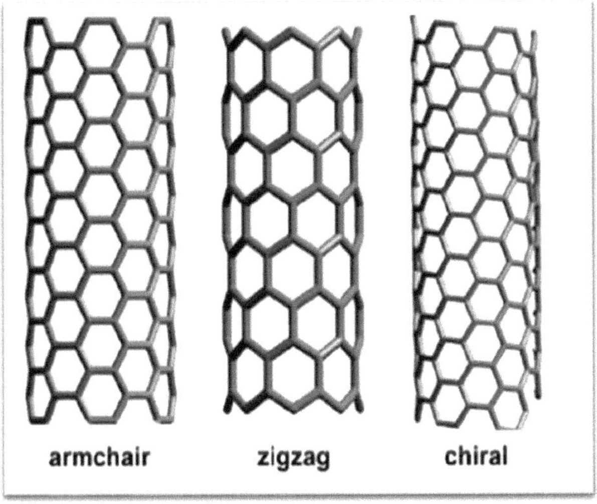


Figure 8: *Structure of a monolayer carbon nanotube as a function of its symmetry.*

CNTs do not give a complete structure and there are errors in the graphene network that makes up a wall [16, 17]. The most common defects are simple atomic defects and Stone-Wales (SW) defects.

An ideal atomic defect (vacant atom) in an armchair (n, n) structure is shown in Figure 9.

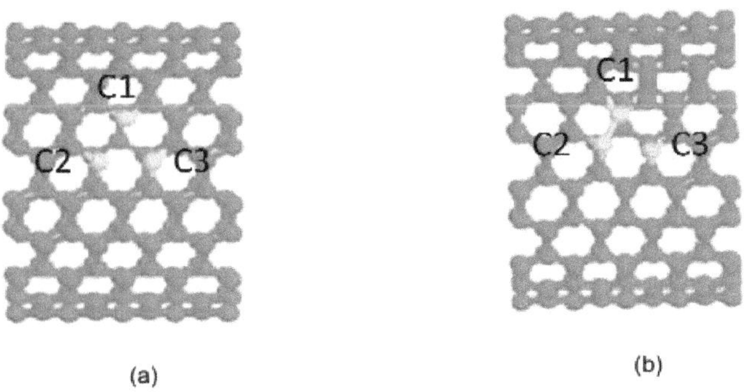


Figure 9: *Configuration of an ideal atomic defect in an (n, n) structure:(a) without a defect and (b) with a vacant atom between the three atoms C1, C2 and C3* [18].

An SW defect can be generated by the rotation of a C-C bond by $\pi$ /2 from its initial site, also making up two pentagons and two heptagons in a local structure. In the case of an (n, n)-type structure, rotation leads to two modified CNT shapes (Figure 10).

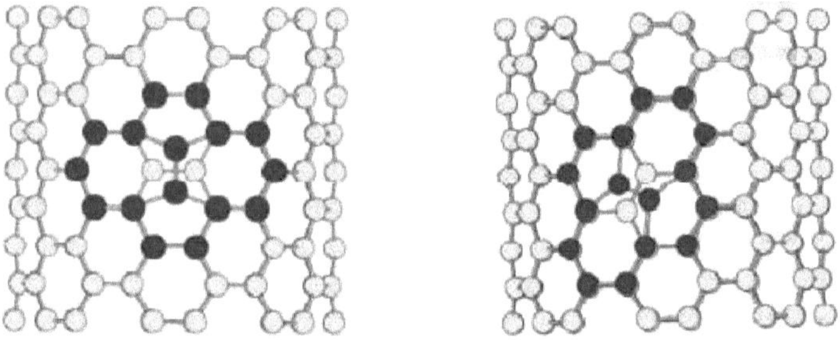

Figure 10: *Configuration of SW defa uts in an (n, n) structure.*

Single atomic defects are more reactive and interact more easily with different kinds of molecules, whereas SW defects are more constant [19]. The presence of these faults influences the physical and chemical properties of CNTs [20, 21] and is the start of plastic deformation of the carbon nanotube.

## II - 3 - Summary

CNTs can be classified into two categories depending on the state of the carbon source. The first technique is based on the sublimation of carbon (graphite), the production of CNT by electric arc [22] and by laser ablation [23] being two examples. The carbon source can be a gas (such as $C_2H_2$, $CH_4$ ...). This is the basis of the second CNT synthesis technique. The most widespread is chemical vapour deposition (CVD). Examples of CNT production using several carbon sources and several synthesis conditions can be found in the literature [24]. Table 1 summarises the techniques mentioned above.

The catalyst-free carbon source is used to produce multi-walled nanotubes (MWCNT), while the use of particles 1 to 3 nm in diameter enables the synthesis of single-walled (SWCNT), double-walled (DWCNT) or triple-walled (TWCNT) CNTs.

Other CNT synthesis techniques have been proposed, such as the use of solar furnaces or low-temperature CNT synthesis. Currently, the CVD synthesis process seems to be the most widely used for the industrial production of CNTs marketed in large quantities.

| Method | Electric arc | Laser Ablation | CVD |
| --- | --- | --- | --- |
| Description | An electric arc generated between two graphite electrodes in the presence of an inert gas. The anode sublimes and forms a plasma. The growth of the CNTs takes place on the cathode. | A graphite target is sublimated by a laser, and the vaporised carbon is carried away by a stream of inert gas. The CNTs are formed on a collector. | CNT is produced by decomposing carbonaceous gas in a furnace at temperatures of around 1200°C. |
| Performance | <75% | <75% | >75% |
| Single or multi-wall NTC | Single and multi-wall | Single and multi-wall | Single and multi-wall |
| Advantage | Simple, low cost | Relatively high purity Well-controlled method. | Simple method, low cost, Temperature Relatively low. Possibility of growing aligned CNTs. High-volume production |
| Disadvantage | Method suitable for producing small quantities of quantities of CNT. The temperatures generated are very high and can reach 6000°C | Adapted method for the production of small quantities of CNT. | The parameters must be finely tuned for the synthesis of SWCNTs or DWCNTs. |

Table 1*: The essentials of CNT synthesis techniques* [25].

## II - 4 - Properties of CNT

The production conditions and different CNT techniques, as well as the different forms of CNT, imply very varied CNT properties. The properties of two individual CNT sources (SWCNT or MWCNT) are shown in Table 2.

| Property | SWCNT | MWCT |
|---|---|---|
| Density (theoretical) | 0,8g/$cm^3$ | 1,8g/$cm^3$ |
| Young's modulus (th & experimental) | 1.3 ± 0.3TPa | 1.28 ±0.6T Pa |
| Electrical conductivity (experimental) | $5.10^3$à$5.10^5$S/m | |
| Specific surface area (estimated) | 1200 g/$m^2$ | 300 g/$m^2$ |

Table 2: *List of some physical properties of individual carbon nanotubes* [26].

Once produced, CNTs require purification or washing [27]. CNT often exist as more or less dense agglomerates (depending on the synthesis and purification technique).

**II - 5 - Carbon nanotube applications**

CNTs can be used primarily for the mechanical reinforcement and electrostatic dissipation of materials. CNTs are used as additives in materials to improve the initial properties of the matrices into which they are introduced. Below are a few non-exhaustive examples of applications.

II - 5 - 1 - Mechanical reinforcement

CNTs can be introduced into epoxy matrices; textile fibres coated with an epoxy matrix reinforced with 0.3 wt.% CNTs give a visible improvement in strength compared with a pure epoxy coating [28].

II - 5 - 2 - Electrostatic dissipation for materials

Nanotubes are used in small quantities in thermoplastic [29-30], thermosetting [31] or elastomeric [32] polymer matrices. Incorporating CNTs makes the matrix more efficient and less sensitive to static electricity.

II - 5 - 3 - Electrostatic dissipation for materials Energy additives

CNTs can be used, for example, to create conductive paper for energy storage devices [33]. The high elasticity and elongated shape of CNTs mean they can be used as AFM microscope tips (Figure 11).

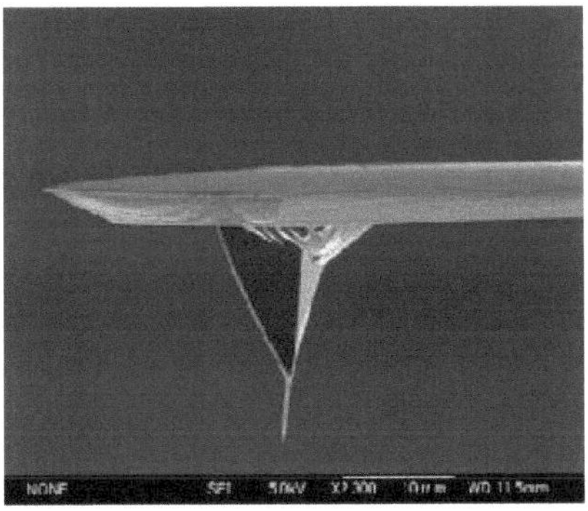

Figure 11: *SEM image of a conventional pyramid-shaped tip to which a bundle of 5 to 10 multilayer nanotubes has been glued* [34].

II - 5 - 4 - In medicine

CNTs can be used for drug delivery [35-36], thanks to functionalities that can make CNTs biocompatible.

CNTs are already present on the industrial market, mainly in the form of "masterbatches". They can be used in certain sports equipment such as tennis rackets, bicycle frames and in prototype flat screens.

## III - NTC PURIFICATION METHODS

CNTs contain various impurities, particularly carbonaceous forms (amorphous carbon, carbon fibres, fullerenes, etc.) and residual metal catalysts. The smaller the diameter of the CNTs produced, the higher the level of impurities [37]. CNT purification methods can be classified into three categories: chemical methods, physical methods and a combination of the two.

### III -1 - Chemical purification methods

The carbon forms normally contained in CNT oxidise more easily than the CNT itself, especially amorphous carbon and fullerene. This is due to the presence of dangling bonds and defects in their structures. Using the principle of selective oxidation, several oxidation techniques have been developed: vapour phase oxidation, liquid phase oxidation and electrochemical oxidation.

### III -2 - Physical purification methods

Physical purification methods have the advantage of preserving the intrinsic structure of the CNTs. Various techniques have been used, such as filtration and centrifugation, and separation by filtration is based on the difference in appearance, shape and solubility between the CNTs, amorphous carbons, fullerenes and metal particles. Small and soluble particles are recovered in the filtrate, while CNTs (with a large shape) remain in the residue. For example, Bonard et al [38] and Shelimov et al [39] used the filtration technique to purify MWCNT and SWCNT respectively. To do this, they first dispersed the CNTs in solution in the presence of SDS by sonication. Once the CNTs were dispersed, they carried out successive filtrations to achieve purities of 90%.

Centrifugation relies on the effect of gravity on the particles and their stability in solution. This difference in stability is linked to the negative charges present on the CNTs previously treated with an acid, as previously shown by Hu et al [40].

The problem with physical methods is the prior dispersion of the CNTs in solution, which requires the addition of a dispersing agent or acid treatment to functionalise the CNTs before purification.

### III -3 - Multi-stage purification

As indicated in the previous paragraphs, a single technique cannot remove all impurities. Several chemical and physical techniques are therefore often combined to obtain pure CNT in high yield.

Bandow et al [41] purified SWCNT by combining microfiltration [42] with gas-phase oxidation (in air). Microfiltration removes soluble particles and fullerenes, followed by treatment at 450°C in air to remove metallic particles. The CNTs obtained are more than 90% pure.

Wang et al. [43] developed a 3-step method to purify, cut and disperse SWCNT synthesised by CCVD: (1) Acid treatment (2.6 M $HNO_3$ at reflux for 24 hours) to remove metal particles, (2) ultrasonication in an acid solution ($HNO_3/H_2SO_4$, $H_2O_2/H_2SO_4$ ) to cut and varnish the SWCNTs (3) high-temperature treatment (1000°C) under $NH_3$ to remove carbon impurities.

The key to multi-stage purification is knowing how to choose the right techniques for the desired end application.

## IV - DISPERSING CARBON NANOTUBES

CNTs need to be scattered in order to take advantage of their properties for several applications. They spontaneously form bundles (Figure 12) by Van der Waals interactions, which can be several microns in size. In order to separate these bundles, it is often necessary to modify the surface of the CNTs.

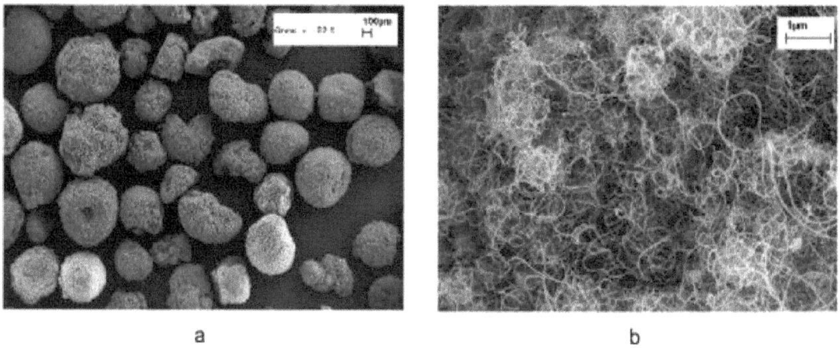

Figure 12: *SEM images of bundles of CNT: (a) low magnification image (b) zoom on one of the bundles* [44].

There are two main approaches to modifying CNT surfaces. The first involves chemically grafting (via covalent bonds) functional groups onto the surface of the CNTs, giving them greater affinity with the solvent. The second involves non-covalent bonds between CNTs and molecules [45, 46].

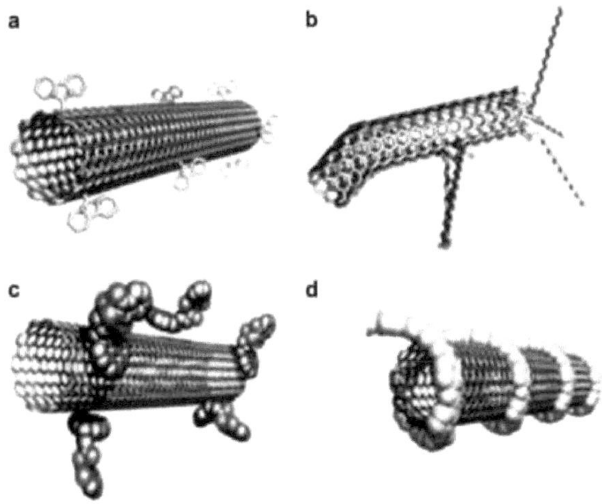


Figure 13: *Schematic representation of functionalized carbon nanotubes using different methods: (a and b) chemical functionalization methods, (c and d) physico-chemical coating method with (c) adsorption of active tension and (d) polymer "wrapping"* [47].

**IV - 1 - Covalent methods**

Carbon nanotubes can be purified and functionalised using acid treatments. Following these treatments, functional groups (acid, carboxyl, alcohol, etc.) appear on CNT surfaces (Figure 14), enabling the subsequent formation of covalent bonds with reactive molecules or functional polymers [48].

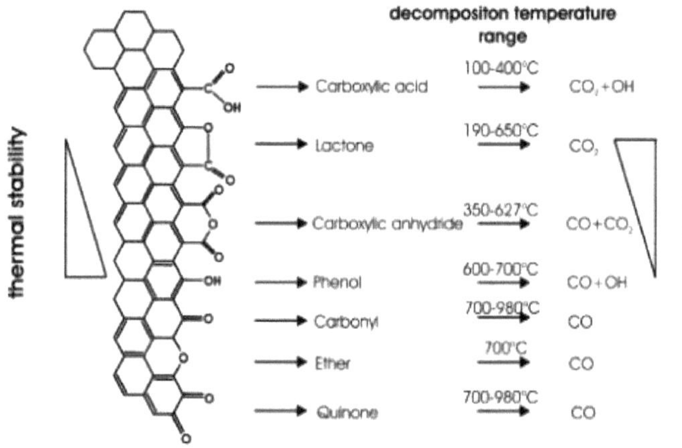


Figure 14: *Different oxygenated functional groups created on the surface of CNT* [49].

There are two grafting methods: grafting to and grafting from.

IV - 1 - 1- The 'grafting to' approach

In this approach, polymers with active functions will reactivate with the functional groups present on the surface of the CNTs. Normally, these functions are carboxyl functions and they are made active by treatment with thyonyl chloride [50] :

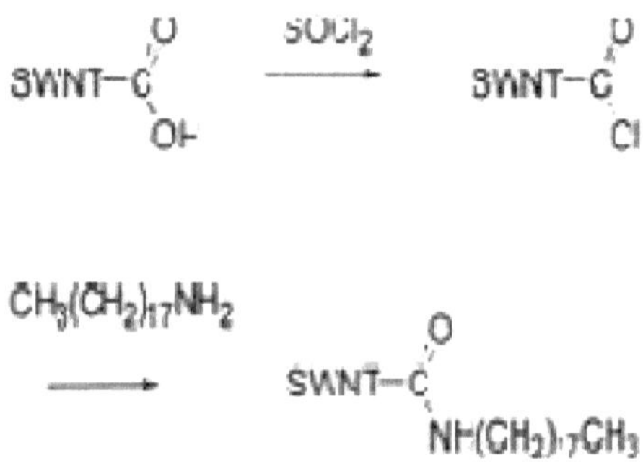


Modified CNTs can thus be covalently bonded with PPE-EI [51], PVOH [52]

(figure 15), POE by an amidation reaction of the terminal amine function [53], glucosamine [54] by the formation of an amide bond between the polymer and the functionalised CNTs, and many other polymers.

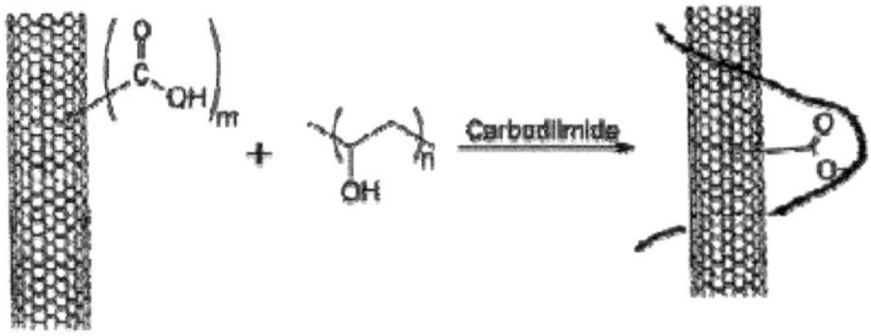


Figure 15: *Esterification reaction between the carboxyl function present on the CNT surface and the polymer to produce a functionalised CNT* [55].

IV - 1 - 2 - The "grafting from" approach

The "grafting from" approach means that polymerisation takes place on the functional group present on the surface of the CNT. This method can be used to manufacture functionalised CNTs while controlling the thickness of the functional layer.

The most commonly used polymerisations are atom transfer radical polymerisation (ATRP) [56, 57] and RAFT [58, 59]. Yan et al [60, 61] synthesised MWCNT-PMMA and MWCNT-PS systems by ATRP using the stepwise procedure shown in Figure 16.

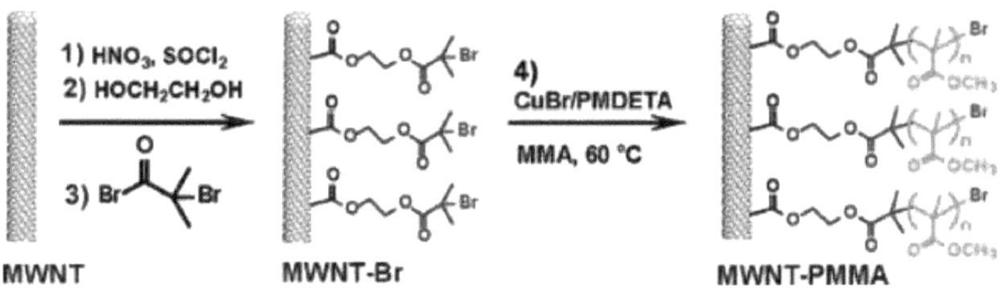


Figure 16: *Functionalisation of MWCNT by ATRP on the surface* [62].

Functionalisation by covalent bonding improves the dispersion of CNTs in solution and, in some cases, reinforces the polymer matrix [63, 64]. However, covalent methods have several disadvantages. Firstly, the grafting process is rather tedious and requires a fairly long total reaction time (up to 70 hours). In addition, covalent functionalisation introduces modifications to the surface of the CNTs, which can affect their mechanical and especially their electrical properties [65-66].

## IV - 2 - Physico-chemical methods

The physico-chemical functionalisation approach is based on the non-covalent interactions that can be established between CNTs and molecules. In particular, these bonds are of the Van der Waals and π-π type [67]. This approach has the advantage of not damaging the CNT structure and of preserving their properties. The problem with this method is that the bonds between CNTs and molecules can be weak.

Researchers distinguish between the phenomenon of adsorption [73, 74] and the phenomenon of "wrapping" molecules around CNTs [68-72]. Wrapping" can be used in the case of polymers around CNTs, whereas adsorption refers to surfactants or small molecules that coat the CNTs [75]. Throughout this manuscript, for ease of understanding, adsorption and wrapping will be considered as equivalent.

### IV - 2 - 1- Surfactants

The surfactant most commonly used for dispersing CNT in aqueous solution is SDS. The state of the dispersion varies with the surfactant concentration [106]. At low SDS concentrations, the system clumps together (figure 19 A, triangles in phase B), and the amount of SDS is not high enough to cover the surface of the CNTs and prevent them from interacting with each other. By increasing the amount of SDS slightly, the CNTs are dispersed homogeneously (Figure 17 A, phase C circles), which can be observed by optical microscopy (Figure 17 C). However, at higher SDS concentrations, depletion interactions appear: the increasingly numerous SDS micelles no longer have room to intercalate between the CNTs already covered with surfactant, which favours attractive interactions between the CNTs.

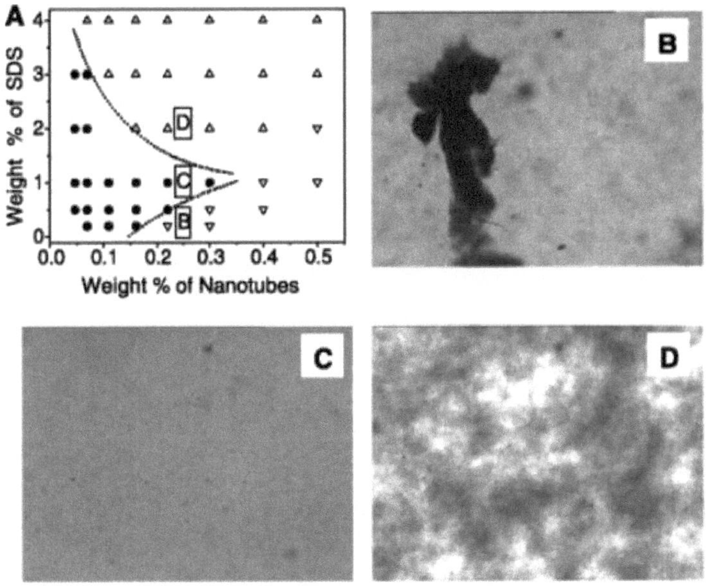


Figure 17: *(A) Phase diagram showing CNT dispersion as a function of SDS concentration; triangles correspond to an aggregated state and circles to a dispersed state. (B-D) Optical microscopy images of different CNT dispersions in solution (horizontal dimension is 300 μm)* [76].

IV - 2 - 2- Homopolymers

A wide variety of polymers have been used to disperse CNT, either in aqueous solution or in the presence of an organic solvent. Smalley et al [99] achieved reversible dispersions of CNT in water using linear polymers of polyvinyl pyrolidone (PVP) or polystyrene sulphate (PSS). They demonstrated that the bond between the polymer and the CNTs is robust, does not depend on the presence of excess polymer in the solution and that the coating is uniform along the length of the tubes.

Other polymers in aqueous solution have also enabled homogeneous dispersion of CNTs, mainly polymetaphenylene vinylene (PmPv) [77] and natural polysaccharides, more specifically gum arabic [78].

Yuan et al. [78] succeeded in synthesising, from 4-[3-(1 Pyrenyl) propylcarbonyloxy]-1-butyne (PyB) and 4-ferrocenylcarbonyloxy-1-butyne (FcB), the corresponding polymers PPyB and PFcB with a fairly high molar mass (up to 33300 g/mol) using a rhodium zwitterionic complex Rh+(nbd)[$C_6H_5B^-$ $(C_6H_5)_3$]. These polymers enable good quality dispersion of the CNTs in various common organic solvents thanks to the electronic interactions between the aromatic nuclei of the polymers and the surface of the CNTs (Figure 18).

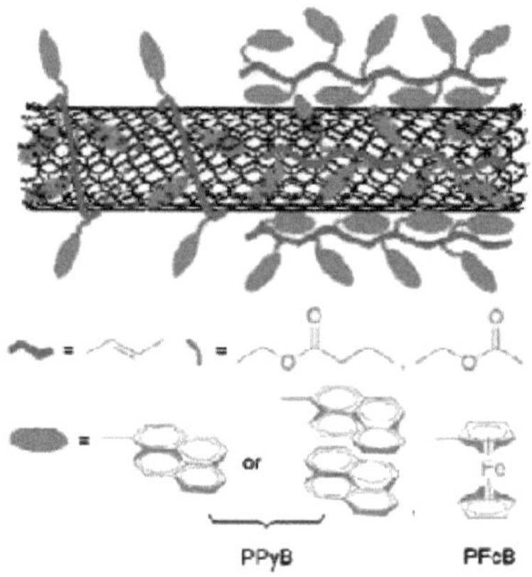


Figure 18: *Illustration showing the interactions between the aromatic nuclei of the polymers (PPyB and PFcB) and the CNT surface* [79].

The use of a polymaleimide (MIP) not only allows homogeneous dispersion of CNTs in different organic solvents, but also improves the physical properties of PPS/NTC-MIP composites compared with those of PPS/NTC composites [80].

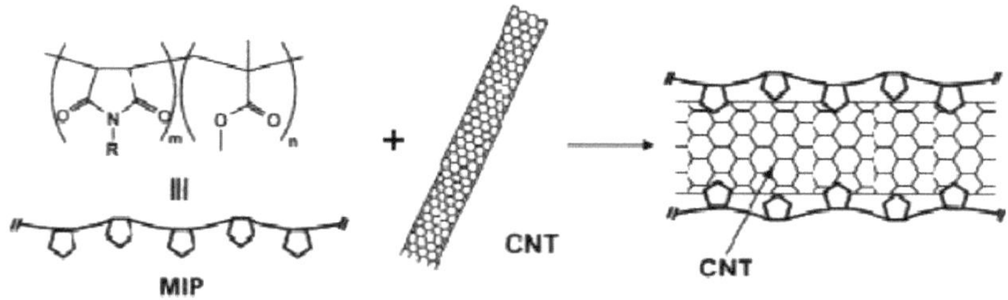


Figure 19: *Illustration showing the adsorption of the MIP polymer onto the CNT surface* [80].

IV - 2 - 3- Peptides and DNA molecules

Carbon nanotubes functionalised with biological molecules have great potential for applications in biotechnology.

Dieckmann et al [81] have used synthetic α-helical amphiphilic peptides to coat CNTs (Figure 20). These systems can be used in electronic circuitry and molecular sensing applications. Kim et al [82] also used peptides, specifically amylose. They succeeded in encapsulating CNTs with amylose in an aqueous solution containing 10 to 20% DMSO.

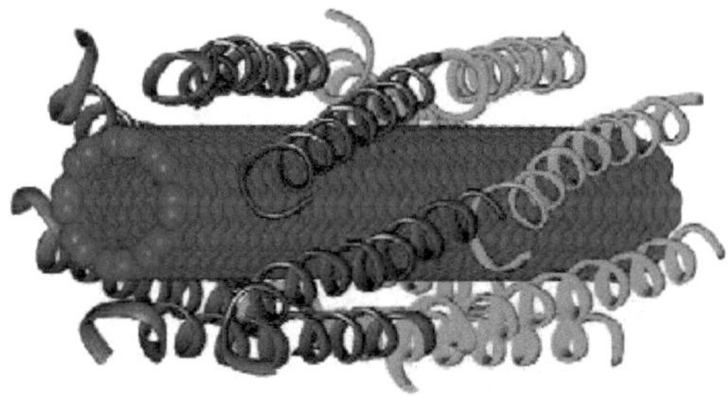

Figure 20: *SWCNT coated with amphiphilic peptides* [82].

Zheng et al [83, 84] were among the first to study the dispersion of CNT in aqueous solution using DNA molecules. They showed that the dispersion of CNT by a DNA strand depends on the DNA sequences that are in contact with the CNT.

IV - 2 - 4 - Block copolymers

Block copolymers are among the most effective steric stabilisers. They are composed of two chemically distinct and often incompatible entities (designated A-B and A-B-A for di-blocks and tri-blocks, respectively) that are covalently bonded [85].

Block copolymers are increasingly used to coat CNTs and stabilise them in solution. A typical scenario for steric stabilisation using block copolymers is based on the dual action of the polymer: while one of the blocks (B) in the chain adsorbs to the surface of the CNTs, the other block (A) expands and repels the other polymer-decorated CNTs. It is well known in the field of colloid science that the determining parameter of the dispersibility of block copolymers is based on solvent selectivity: a selective solvent that acts as a 'good solvent' for one of the blocks (e.g. block A) dissolves the polymer easily and increases the spatial dimensions of the polymer chain, while acting as a 'bad solvent' for the other block (B) which adsorbs onto the surface of the CNTs.

Since CNTs are hydrophobic, amphiphilic block copolymers in aqueous solution are one of the most interesting systems.

IV - 2 - 5 - In organic solvents

Zou et al [86] succeeded in dispersing CNTs in chloroform using conjugated polymers $P_3$HT-PS conjugated polymers (34-65 wt%). They demonstrated that the $P_3$HT block comes into contact with the CNTs by п-п interactions whereas the PS block allows stability in solution. They then used these dispersions to produce PS composites.

A similar result was obtained with polyethylene oxide -(hexa-p-phenylene)-polyethylene oxide tri-block copolymers [87].

Shvartzman-Cohen et al [88] have shown that the nature of the solvent in which CNTs are placed in the presence of block copolymers influences dispersion. In the presence of a good solvent for both blocks, dispersion will not occur (Figure 21, solution F).

Figure 21: *Dispersions of SWNT in solution in the presence of block copolymers: 1 wt % SWNT in 1 wt % PS-tbuAC solution (A) in ethanol and (B) in 2-propanol. (C) 1 wt % POE-PDMS-PEO in heptane, (D) 1 wt % POE-PPO-POE (F-88) in water, (E) 1 wt % PPO-POE (PE10, 500) in water, and (F) 1 wt % POE-PDMS-POE in toluene. Images taken two months after preparation* [88].

Sluzarenko et al [89] developed a flexible method for dispersion and stabilisation of nanotubes using polystyrene-b polyisoprene block copolymers in different organic solvents. Stable dispersions were obtained in polar (DMF) and apolar (heptane) solvents depending on the selectivity of the diblock copolymers. They showed that the direct interaction between the nanotubes and the different blocks is not the most important parameter for stabilisation. The major factor seems to be the selectivity of the solvent towards the block copolymer and the block copolymer ratio. The best stabilisation was obtained when the solvent was good for the largest block.

IV - 2 - 6 - In aqueous solution

Agarwal et al [90] obtained good dispersions of SWNT in water using PS-POE (polystyrene-polyethylene oxide) and PMMA-POE (poly methyl methacrylate-polyethylene oxide) block copolymers. They used copolymers with low molar masses (1000 and 3000 g/mol respectively). However, it was not possible to disperse SWNTs with these two copolymers in toluene (toluene being a good solvent for the hydrophobic PS and PMMA parts). This shows that the solubilisation of the SWNTs in water is made possible by the physical associations between the hydrophobic blocks and the CNTs, while the hydrophilic block expands in solution and thus allows good dispersion of the SWNTs by steric stabilisation. The same group also demonstrated the effect of sonication mode and time on the length and diameter of the SWNTs, by monitoring changes in the viscosity of the dispersions.

Ho Jo et al [91] synthesised and used oligothiophene polyethylene glycol (TN-PEG) copolymers with two lengths of oligothiophene (3TN-PEG and 5 TN-PEG) for CNT dispersion in aqueous media. Significant adsorption of the thiophene block was observed via л- л interactions. CNT dispersion was better with long thiophene chains and the electrical conductivity of TN-PEG/NTC films was greater than that obtained with non-conductive polymers such as polyethylene oxide-polypropylene oxide tri-block copolymers (pluronic F127).

Kim et al [92] succeeded in encapsulating SWNTs in aqueous solution using an amphiphilic PEtOz-PCL block copolymer, formed from a hydrophilic poly (2-ethyl-2-oxazoline) part and a hydrophobic polycaprolactone part. The PEtOz-PCL/SWNT complex disperses well in water, with the PCL part adsorbed on the SWNTs and the PEtOz part in solution in water.

As part of the preparation of CNT/nanohybrid polyelectrolyte with an amphiphilic block, for use in the synthesis of semiconducting CdS nanoparticles, N. Tagmatararchis et al [93] used an amphiphilic PSHI block copolymer (formed from a polystyrene block and a sodium(2-sulphomate-3-carboxylate) isoprene block). Aqueous CNT/PSHI solutions are stable for several months without precipitation.

Finally, tri-block POE-PPO-POE polymers (Pluronic® F127 with a molecular weight of12600g/mol) have also been used to stabilise aqueous CNT solutions [92, 93].

The use of amphiphilic block copolymers is becoming increasingly important in

CNT dispersion for several reasons:

-The copolymers encapsulate the CNTs (non-covalent bonding). This means that the structure and properties of the CNTs are not impaired.

-The possibility of taking advantage of the bi-functionality of the blocks and the different properties that each of them can bring.

-The presence of the hydrophobic block enables better adsorption on the surface of the CNT (the surface of CNT is hydrophobic).

-Dispersions in aqueous solution are easy to handle and characterise.

# V - NANOCOMPOSITES

## V - 1 - Differences between composite and nanocomposite

A composite is a solid phase material made up of at least two components whose respective qualities complement each other to form a material with improved overall performance. For example, if a naturally insulating plastic is placed with a conductive metal in laminated layers, the result is an anisotropic material with conductive properties parallel to the layers and insulating properties perpendicular to the layers [94]. Composite materials contain inhomogeneities, but these are small enough for the materials to remain homogeneous on a macroscopic scale. Composite materials exist naturally, such as wood, which is made up of lignin into which cellulose fibres are embedded.

What is different about nanocomposites is the size of the fillers present in the matrix. They are of the order of 10 nm in size and are present in different forms: one-dimensional (CNT), two-dimensional (graphene and clay) or three-dimensional (nanosphere). Bone is an example of a nanocomposite.

Since the fillers added to the matrices are small in size, the interfacial surface area is larger, resulting in a significant improvement in properties compared with conventional composites. In addition, the filler content is up to 20 times lower than that used for macroscopic fillers.

| | Carbon fibre | Carbon nanotube |
|---|---|---|
| Particle size | 10μm | 10nm |
| Percentage by volume | 30% | 3% |
| Number of particles | $10^5$ | $10^{15}$ |
| Interface | ~0 ,1$m^2$ | ~100$m^2$ |
| Form ratio | ~100 | >1000 |

Table 3: *Comparison of the characteristics of macroscopic particles (carbon fibres) and nanometric particles (carbon nanotubes).*

However, the addition of fillers has disadvantages that affect the development and properties of the final composite:

-Increase in system viscosity.

-Difficulties in dispersing charges in the matrix.

Black colouration of the matrix in the presence of carbon-based particles.

-Weak matrix-charge cohesion.

-Accounting required.

There are two main types of composite: nanocomposites based on a metallic or ceramic (mineral) matrix and composites based on a polymer (organic) matrix.

## V - 2 - Polymer-carbon nanotube composites

### V - 2 - 1- Preparation methods

Various development strategies have been developed, but in order to obtain high-performance materials with excellent electrical and mechanical properties, the development method is crucial. The most commonly used techniques are solution dispersion [96-100], melt dispersion [129, 130] and in situ polymerisation in the presence of CNTs [101, 102].

| Method of preparation | Principle | Benefits | Disadvantages |
|---|---|---|---|
| Mix in solution | Dispersion of CNTs and polymer in aqueous solution aided by ultrasound and/or CNT functionalisation. | The use of low levels of CNT. | Use and disposal of solvents. |
| Mix with In the molten state | Extrusion and injection are the most commonly used methods for obtaining better dispersions. Functionalisation and accounting for CNTs can also be used. | Easy to produce to scale industrial (simple, clean, solvent-free). | Viscosity in the system. |
| In situ polymerisation | There are two types of in situ polymerisation :<br>- Conventional polymerisation in the presence of CNTs (in this case there is no covalent bond with the CNTs).<br>- Grafting from (see paragraph I.2.1). | In the case of "grafting from", the properties mechanical properties of composites are improved. | Technical generally long. |

Table 3: *Comparison of the different methods for producing polymer/NTC composites.*

### V - 2 - 2- The percolation phenomenon

Percolation is a communication problem that arises in a large environment in which a large number of "sites" likely to relay information locally are regularly distributed.

These sites communicate with each other via links whose effectiveness is random. Depending on whether or not the proportion of active links exceeds a threshold value, it may or may not be possible to transmit information over long distances. (according to Encycloedia University). In the case of a composite, the percolation threshold therefore characterises the concentration of one of the phases above which percolation occurs, manifesting itself as a sudden change in the properties observed.

In the case of CNT-polymer composites, the percolation threshold will enable us to determine the quantity of CNT we need for them to percolate together and create a network that will enable us to reveal the specific properties of the CNTs.

Depending on the type of percolation studied, the notion of contact between the CNTs is different. Electron transfer (electrical percolation) does not take place in the same way as stress transfer (mechanical percolation). We therefore observe different percolation thresholds depending on the properties being studied.

V - 2 - 3 - Theoretical approach to the percolation threshold

The concept of excluded volume $<V_{ex}>$ has proven to be a powerful method for estimating the percolation threshold of composites containing statistically non-spherical dispersed particles [103], such as CNTs. The concept is based on the idea that the percolation threshold is not related to the actual volume of particles, but rather to their $V_{ex}$ . the latter is defined as the volume around an object into which the centre of another object of similar shape is not allowed to penetrate [104].

The volume of a cylinder is given by the following formula:

$$V = \frac{D2}{4} L\pi \ (1)$$

Where D and L are the diameter and length of the cylinder respectively.

In the case of CNTs, the form factor μ:

$$\mu = L/D(2)$$

From formulae (1) and (2) we can determine the steric exclusion volume in the case of CNTs:

$$< Vex >\approx \frac{DL2}{2}\pi(3)$$

And the percolation threshold $p_c$ can be calculated as [105]:

$$Pc{\sim}v/Vex = D/2L = 1/2\mu(4)$$

In the case of CNTs, the form factor can vary between 100 and 10,000 depending on the synthesis method and the treatments the CNTs have undergone (sonification, acid treatment, etc.). The percolation threshold is then between 0.01 wt.% (for the longest CNTs) and 1 to 2 wt.% (for the shortest CNTs). These values largely correspond to those used in the literature for different polymer matrices. For example, 0.5 wt.% can be found for PMMA [106] and 2 wt.% for PE [107].

The calculation presented earlier treats the CNTs as rigid tubes. However, TEM images show CNTs as "undulating" objects. The effect of CNT undulation on the percolation threshold has been discussed in several publications using different theoretical approaches [108, 109]. All show that the percolation threshold increases with increasing CNT "waviness" (Figure 22). However, in all the calculations, the increase in $p_c$ remains well below a factor of 2, which means that the effect of the corrugation can be considered negligible compared with the effect of the steric exclusion volume.

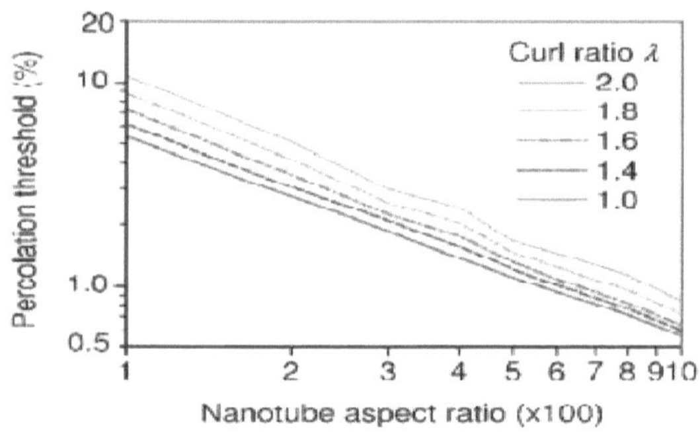


Figure 22: *Simulation showing the effect of CNT undulation on the percolation threshold* [109].

The effect of the presence of aggregates on the percolation threshold was also studied by Grujicic et al [110]. As expected, the percolation threshold increases as the size of the aggregates present in the composite increases.

There is a percolation law which describes the change in behaviour around the percolation threshold [111, 112]. This law is presented in equation (5)

$$P(p) = (p - pc)t (5)$$

With P(p) the probability that, for a certain fraction of occupied sites p, a site belongs to the percolating cluster. In practice, P(p) may correspond to the conductivity of the system or to the relative viscosity measured, p being the volume or mass fraction of the fillers, $p_c$ the critical concentration or percolation threshold and t the critical exponent [113]. Some authors relate the variation of the t factor to the dimension of the system, with calculated values of t ~1.33 corresponding to a two-dimensional system and t ~2 to a three-dimensional system [114]. However, it seems rather complicated to extract precise geometrical information about the CNT network from experimentally determined t values. Such a procedure is generally not justified.

Studies have shown a difference between the electrical percolation threshold and the rheological percolation threshold. This is because charged particles (such as CNTs) can interact with each other via polymer chains, whereas direct contact is necessary for electrical conduction. For the SWCNT / PMMA thermoplastic system, Du et al [114] observed a rheological threshold at 0.12 wt.%, below the electrical threshold at 0.39 wt.%. Different behaviour was observed for MWCNT in a thermoset system.

Due to the absence of polymer chains in the "liquid" state (which can be disentangled) in thermoset systems, the electrical percolation threshold is reached first (at around 0.04 wt.%) whereas rheological percolation is reached at 0.1 wt.% [115] .

Various polymer matrices have been used to disperse CNTs, which can be classified into two main groups: thermosets and thermoplastics.

### V - 2 - 4 - Thermosetting composites

The thermosetting composites that have been studied the most are epoxy composites because of their important applications in various fields. Epoxy/NTC composites have been prepared either by the melt route or by the solvent route. Ajayan et al [129] were among the first to prepare epoxy matrices with MWCNTs. The composites were prepared by mechanical mixing and the orientation was observed on sections with a thickness of less than 200 nm.

Jin et al [116] also produced epoxy-based composites with CNTs aligned within. The composites were prepared by solvent (in chloroform) and uniaxially stretched at 100°C. These authors showed that the CNTs remained aligned even after the filler was removed at room temperature (Figure 23).

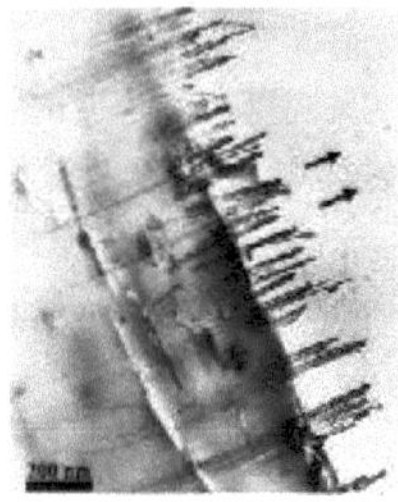

Figure 23: *TEM image of an internal fracture in the epoxy composite after being cut parallel to the tensile direction. The thickness of the sample is 90 nm. The CNTs are aligned parallel to the tensile direction* [116].

In order to improve the dispersion of CNT in the epoxy matrix, treatments of CNT with surfactants have been developed [116, 117].

All these studies show a shift in the $T_g$ towards higher values and an improvement in the elastic modulus of composites.

### V - 2 - 5 - Thermoplastic composites

A number of composites based on CNTs and thermoplastic matrices have been developed by different teams. In this section, we will present just a few examples. The major problem with all these studies is the difficulty of comparing them, mainly because of the nature of the CNTs chosen, the matrix used and the techniques for preparing and dispersing the CNTs in the matrix.

### V - 2 - 6 - PMMA-based composites

Table 4 shows the different preparation routes for some PMMA/CNT composites presented in the literature and the main characteristics observed.

| Preparation methods | By solvent | By mixing in the molten state | By emulsion polymerisation in aqueous solution in the presence of SDS. |
|---|---|---|---|
| Comments | Percolation threshold | An improvement in elastic | Functionalization of CNT by SDS leads |

| | | | |
|---|---|---|---|
| | rheology (0.12 wt.%) lower than the electrical percolation threshold<br>electrical percolation threshold (0.39 wt.%) | modulus (Figure 24) | to a decrease in the rheological percolation threshold (from 2.5 to 1.5 wt.%) |

Table 4 - Methods of making PMMA-based composites and observations obtained [114, 118-190,120,123].

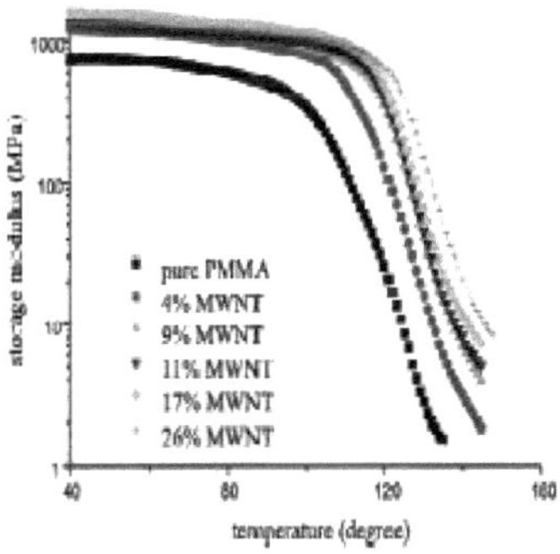


Figure 24: *Elastic modulus as a function of temperature for different PMMA-based composites* [120].

V - 2 - 7 - PE-based composites

Valentino et al [124] show close rheological and electrical percolation threshold values of between 1 and 2.5 wt.% (figure 25) and a better dispersion of CNTs in the case of high-density polyethylene compared with low-density polyethylene [154], probably because the mixing temperature was not suitably chosen, this is highlighted in the SEM image (figure 26) of low-density PE + 2.5 wt.% MWNT, which clearly shows dark areas with a low concentration of nanotubes and light areas with a high concentration of CNTs.

In contrast, McNally et al [125] found rheological and electrical percolation threshold values close to and much higher (7.5 wt.%) than the previous study. This difference can be explained by differences in composite preparation. In both cases, melt preparation using a twin-screw extruder was used, but the residence time increased from 2 to 10 min and the rotation speed doubled in value (from 25 to 50 rpm) between the second and first study.

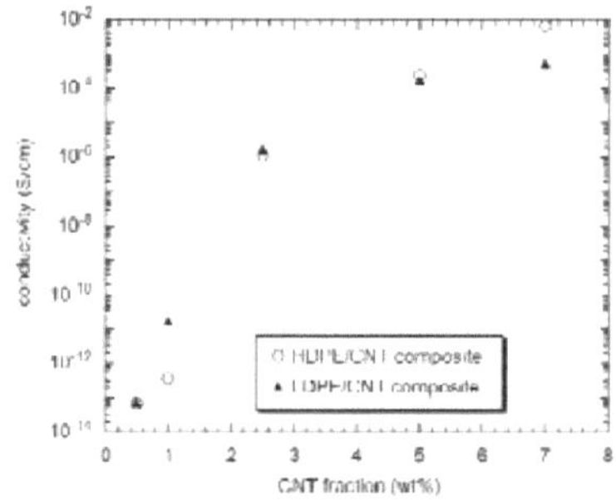


Figure 25: *Conductivity as a function of CNT mass concentration for low density (LDPE) and high density (HDPE) PE matrices* [126].

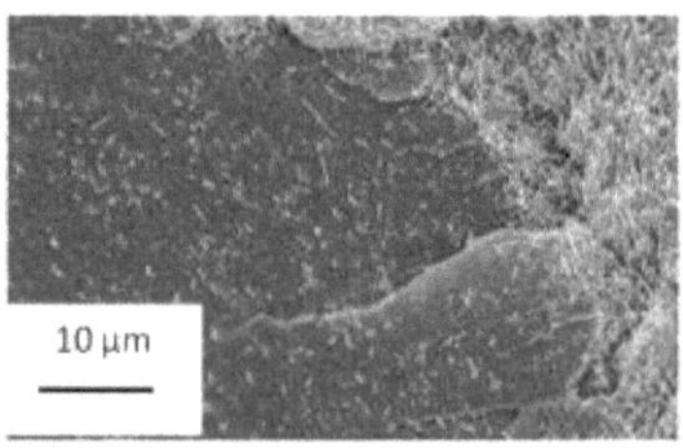


Figure 26: *SEM image of low density PE + 2.5 wt.% MWNT* [127].

V - 2 - 8 - <u>POE-based composites</u>

Abraham et al [128] and Chatterjee et al [129] produced POE/NTC nanocompsosites in aqueous solution, with MWCNT and SWCNT respectively. Rheological and electrical studies were carried out on samples with different CNT contents in order to determine the electrical and rheological percolation thresholds. The variation in crystallinity as a function of CNT content was also analysed. The results are shown in Table 5.

| | SWCNT | MWCNT |
|---|---|---|
| Rheological analysis | Percolation threshold rheological at 0.09 wt.%. | The two rheological and electrical studies showed the existence of a single percolation threshold at around 1 wt.%. |
| Electrical analysis | Percolation threshold electric at 0.03 wt.%. | |
| Crystallinity studies | The degree of crystallinity decreases with increasing CNT concentration | The level of crystallinity peaks at low levels of CNT before decreasing at higher levels (Figure 27). |
| Remarks | SWCNT functionalized with LDS | Non-functionalised MWCNT |

Table 5: *Main results of characterisations (rheological, electrical and thermal) on POE/NTC composites.*

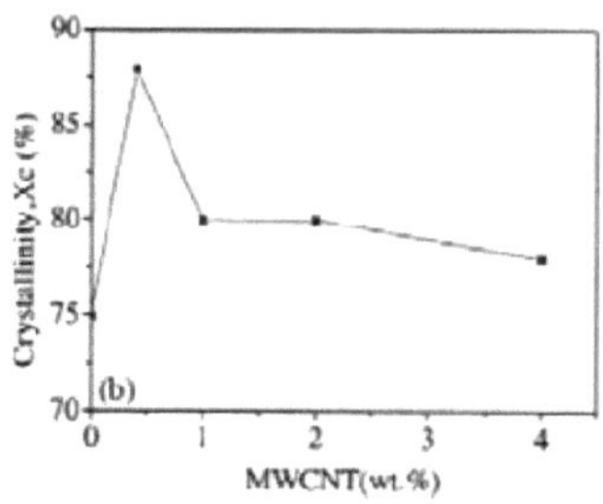


Figure 27: *DSC crystallinity results for POE composites with different MWCNT concentrations* [128].

V - 2 - 9 - Composites with different thermoplastic polymers

CNTs have been produced by several groups [130-132] for photovoltaic applications. The films were prepared using a solvent process and electrical and optical studies showed an improvement in the electrical and optical properties in the presence of CNTs. The films are also more mechanically resistant.

PVA matrices have also been used to disperse CNTs by the aqueous route [133-135]. The presence of 1 wt.% CNT increases the nucleation rate of the PVA.

Wong et al [136] produced polystyrene/CNT films using a solvent process. TEM observations show good adhesion between the matrix and the CNTs. Mechanical tests carried out on samples with CNT content ranging from 0.1 to 2 wt.% showed a slight increase in strain rate at break and elastic modulus compared with the polymer alone.

Andrews et al [136] prepared polypropylene/CNT and polystyrene/CNT films by the solvent route. They showed that the electrical percolation threshold depended on the matrix in which the CNTs were dispersed. The threshold value is 0.05 vol.% in the case of polypropylene, but this value increases in the case of polystyrene. They also noted a slight improvement in mechanical properties when CNTs were incorporated.

To conclude this chapter, it should be emphasised that the results obtained for polymer/CNT composites are highly inconsistent, depending on the CNT used and the dispersion and characterisation techniques employed. This lack of homogeneity in the different studies carried out by several groups makes it difficult to classify polymer/CNT composites.

## References


[01] H. O. Pierson, Handbook of carbon, graphite, diamond and fullerenes: properties, processing and applications, 1993, Noyes Publications, Park Ridge.
[02] B. I. Yakobson, R. E. Smalley, American Scientist, 1997, 85, 324.
[03] Nanotubes from carbon, P. M. Ajayan, Chem. Rev, 1999, 99, 1787-1799.
[04] http://cnst.rice.edu/pics.html
[05]Kroto HW, Heath JR, O'Brien SC, Curl RF, Smalley RE. C60: Buckminsterfullerene. Nature.1985;318(6042):162-3.
[06]Hillert M, Lange, N. . the structure of graphite filaments. Zeitschrift fur Kristallographie. 1958;111(111):24.
[07]Oberlin A, Endo M, Koyama T. Filamentous growth of carbon through benzene decomposition.Journal of Crystal Growth. 1976;32(3):335-49.
[08]Iijima S, Ichihashi T. Erratum: Single-shell carbon nanotubes of 1-nm diameter (Nature (1993)363 (603-605)). Nature. 1993;364(6439):737.
[09] Bethune DS, Kiang CH, De Vries MS, Gorman G, Savoy R, Vazquez J, et al. Cobalt-catalysed growth of carbon nanotubes with single-atomic-layer walls. Nature. 1993;363(6430):605-7.
[10] Journet C, Maser WK, Bernier P, Loiseau A, Lamy de la Chapelle M, Lefrant S, et al. Large-scale production of single-walled carbon nanotubes by the electric-arc technique. Nature. 1997;388(6644):756-8.
[11]Thess A, Lee R, Nikolaev P, Dai H, Petit P, Robert J, et al. Crystalline ropes of metallic carbonanotubes. Science. 1996;273(5274):483-7.
[12]De Jong KP, Geus JW. Carbon Nanofibers: Catalytic Synthesis and Applications. Catalysis Reviews- Science and Engineering. 2000;42(4):481-510.
[13] Iijima S, Yudasaka M, Yamada R, Bandow S, Suenaga K, Kokai F, et al. Nano-aggregates of single-walled graphitic carbon nano-horns. Chemical Physics Letters. 1999;309(3-4):165-70.
[14] Wagner RS, Ellis WC, Jackson KA, Arnold SM. Study of the filamentary growth of silicon crystals from the vapor. Journal of Applied Physics. 1964;35(10):2993-3000.
[15] Mawhinney DB, Naumenko V, Kuznetsova A, Yates JT, Liu J, Smalley RE. Surface defect site density on single walled carbon nanotubes by titration. Chemical Physics Letters. 2000;324(1-3):213-6. [13] Volpe M, Cleri F. Role of surface chemistry in hydrogen adsorption in single-walled carbon nanotubes. Chemical Physics Letters. 2003;371(3-4):476-82.
[16]He H-y, Pan B-c. Studies on structural defects in carbon nanotubes. Frontiers of Physics in China.2009;4(3):297-306.
[17]Chico L , Crespi VH ,Benedict LX, Louie SG, Cohen ML. Pure Carbon Nanoscale Devices: NanotubeHeterojunctions.PhysicalReviewLetters.1996;76(6):971.
[18] Nardelli MB, Yakobson BI, Bernholc J. Mechanism of strain release in carbon nanotubes. Physical Review B - Condensed Matter and Materials Physics. 1998;57(8).
[19]Rubio A. Spectroscopic properties and STM images of carbon nanotubes. Applied Physics A:Materials Science & Processing. 1999;68(3):275-82
[20]S. Iijima, Nature, vol.354, pp.56-58, 1991, doi:10.1038/354056a0
[21] T. Guo, P. Nicolaev, A. Thess, D. T. Colbert, R.E Smalley, Chem. Phys. Lett. vol. 243, pp. 49-54. Retrieved February 13, 1995.
[22] M. J-Yacamán, H. Terrones, L. Rendón and J. M. Domínguez, Carbon, vol.33, Issue 5, pp.669-678, 1995.
[23] C. E. Baddour, C. Briens, International Journal of Chemical Reactor engineering, vol.3, R3, 2005 .

[24] Xiao-Lin Xie, Yiu-Wing Mai and Xing-Ping Zhou, Materials Science and Engineering: R: Reports, vol.49, Issue 4, pp. 89-112, 2005.
[25] T. W. Ebbesen, H.J. Lezec, H. Hiura, J. W. Bennett, H. F. Ghaemi and T.Thio, Nature, vol. 382, 1996
[26] Siddiqui NA, Sham ML, Tang BZ, Munir A, Kim JK. Tensile strength of glass fibres with carbon nanotube-epoxy nanocomposite coating. Composites Part A: Applied Science and Manufacturing. 2009;40(10):1606-14.
[27] Mi M, ccaron, uscaroník, Omastová M, Krupa I, Prokescaron J, et al. A comparative study on the electrical and mechanical behaviour of multi-walled carbon nanotube composites prepared by diluting a masterbatch with various types of polypropylenes. Journal of Applied Polymer Science. 2009;113(4):2536-51.
[28] Ciardelli F, Coiai S, Passaglia E, Pucci A, Ruggeri G. Nanocomposites based on polyolefins and functional thermoplastic materials. Polymer International. 2008;57(6):805-36.
[29] Ahn BW, Chi YS, Kang TJ. Preparation and characterization of multi-walled carbon nanotube/poly(ethylene terephthalate) nanoweb. Journal of Applied Polymer Science. 2008;110(6):4055-63.
[30]Yu A, Hu H, Bekyarova E, Itkis ME, Gao J, Zhao B, et al. Incorporation of highly dispersed single-walled carbon nanotubes in a polyimide matrix. Composites Science and Technology. 2006;66(9):1187-94.
[31] Fletcher A, Gupta MC, Dudley KL, Vedeler E. Elastomer foamnanocomposites for electromagnetic dissipation and shielding applications. Composites Science and Technology. 2010;70(6):953-8.
[32] Hu L, Choi JW, Yang Y, Jeong S, La Mantia F, Cui LF, et al. Highly conductive paper for energy-storage devices. Proceedings of the National Academy of Sciences of the United States of America. 2009;106(51):21490-4.
[34]Dai H, Hafner JH, Rinzler AG, Colbert DT, Smalley RE. Nanotubes as nanoprobes in scanning probe microscopy. Nature. 1996;384(6605):147-50.
[35] Tran PA, Zhang L, Webster TJ. Carbon nanofibers and carbon nanotubes in regenerative medicine. Advanced Drug Delivery Reviews. 2009;61(12):1097-114.
[36]Bianco A, Kostarelos K, Partidos CD, Prato M. Biomedical applications of functionalised carbonnanotubes. Chemical Communications. 2005(5):571-7.
[37] Hou P-X, Liu C, Cheng H-M. Purification of carbon nanotubes. Carbon. 2008;46(15):2003-25.
[38] Bonard J-M, Stora T, Salvetat J-P, Maier F, Stöckli T, Duschl C, et al. Purification and size-selection of carbon nanotubes. Advanced Materials. 1997;9(10):827-31. [39] Shelimov KB, Esenaliev RO, Rinzler AG, Huffman CB, Smalley RE. Purification of single-wall carbon nanotubes by ultrasonically assisted filtration. Chemical Physics Letters. 1998;282(5-6):429-34.
[40] Hu H, Yu A, Kim E, Zhao B, Itkis ME, Bekyarova E, et al. Influence of the Zeta Potential on the Dispersability and Purification of Single-Walled Carbon Nanotubes. The Journal of Physical Chemistry B. 2005;109(23):11520-4.
[41] Bandow S, Asaka S, Zhao X, Ando Y. Purification and magnetic properties of carbon nanotubes. Applied Physics A: Materials Science & Processing. 1998;67(1):23-7.
[42] Bandow S, Rao AM, Williams KA, Thess A, Smalley RE, Eklund PC. Purification of Single-Wall Carbon Nanotubes by Microfiltration. The Journal of Physical Chemistry B. 1997;101(44):8839-42.
[43] Wang Y, Gao L, Sun J, Liu Y, Zheng S, Kajiura H, et al. An integrated route for purification, cutting and dispersion of single-walled carbon nanotubes. Chemical Physics Letters. 2006;432(1-3):205-8.
[44] S. Bordere JMC, NE. El Bounia, P. Gaillard, N. Passade-Boupat, P.M. Piccione, D. Plée.

INDUSTRIAL PRODUCTION AND APPLICATIONS OF CARBON NANOTUBES. wwwgraphistrengthcom
[45] Park C, Lee S, Lee JH, Lim J, Lee SC, Park M, et al. Controlled assembly of carbon nanotubes encapsulated with amphiphilic block copolymer. Carbon. 2007;45(10):2072-8.
[46] Mountrichas G, Pispas S, Tagmatarchis N. Aqueous Carbon-Nanotube-Amphiphilic-Block-Copolymer Nanoensembles: Towards Realization of Charge-Transfer Processes with Semiconductor Quantum Dots13. Small. 2007;3(3):404-7.
[47] Hirsch A. Functionalization of single-walled carbon nanotubes. Angewandte Chemie - International Edition. 2002;41(11):1853-9.
[48] Kitano H, Tachimoto K, Anraku Y. Functionalization of single-walled carbon nanotube by the covalent modification with polymer chains. Journal of Colloid and Interface Science. 2007;306(1):28-33.
[49] Muckenhuber H, Grothe H. The heterogeneous reaction between soot and NO2 at elevated temperature. Carbon. 2006;44(3):546-59.
[50]Chen J, Hamon MA, Hu H, Chen Y, Rao AM, Eklund PC, et al. Solution Properties of Single-Walled Carbon Nanotubes. Science. 1998 October 2, 1998;282(5386):95-8.
[51] Riggs JE, Guo Z, Carroll DL, Sun Y-P. Strong Luminescence of Solubilized Carbon Nanotubes. Journal of the American Chemical Society. 2000;122(24):5879-80.
[52] Lin Y, Zhou B, Shiral Fernando KA, Liu P, Allard LF, Sun Y-P. Polymeric Carbon Nanocomposites from Carbon Nanotubes Functionalized with Matrix Polymer. Macromolecules.2003;36(19):7199-204.
[53] Sano M, Kamino A, Okamura J, Shinkai S. Self-Organization of PEO-graft-Single-Walled Carbon Nanotubes in Solutions and Langmuir-Blodgett Films. Langmuir. 2001;17(17):5125-8.
[54] Pompeo F, Resasco DE. Water Solubilization of Single-Walled Carbon Nanotubes by Functionalization with Glucosamine. Nano Letters.2002;2(4): 36973.
[55] Liu Y, Yao Z, Adronov A. Functionalization of Single-Walled Carbon Nanotubes with Well-Defined Polymers by Radical Coupling. Macromolecules. 2005;38(4):1172-9.
[56] Shanmugharaj AM, Bae JH, Nayak RR, Ryu SH. Preparation of poly(styrene-co-acrylonitrile)-grafted multiwalled carbon nanotubes via surface-initiated atom transfer radical polymerization. Journal of Polymer Science Part A: Polymer Chemistry. 2007;45(3):460-70.
[57] Qin S, Qin D, Ford WT, Resasco DE, Herrera JE. Polymer Brushes on Single-Walled Carbon Nanotubes by Atom Transfer Radical Polymerization of n-Butyl Methacrylate. Journal of the American Chemical Society. 2003;126(1):170-6.
[58]Ye-Zi Y, et al. Directly growing ionic polymers on multi-walled carbon nanotubes via surface RAFT polymerization. Nanotechnology. 2006;17(9):2350.
[59] Hong C-Y, You Y-Z, Pan C-Y. A new approach to functionalize multi-walled carbon nanotubes by the use of functional polymers. Polymer. 2006;47(12):4300-9.
[60] Kong H, Gao C, Yan D. Controlled Functionalization of Multiwalled Carbon Nanotubes by in Situ Atom Transfer Radical Polymerization. Journal of the American Chemical Society. 2003;126(2):412-3.
[61] Kong H, Gao C, Yan D. Functionalization of Multiwalled Carbon Nanotubes by Atom Transfer Radical Polymerization and Defunctionalization of the Products. Macromolecules. 2004;37(11):4022-30.
[62] Blake R, Gun'ko YK, Coleman J, Cadek M, Fonseca A, Nagy JB, et al. A Generic Organometallic Approach toward Ultra-Strong Carbon Nanotube Polymer Composites. Journal of the American Chemical Society. 2004;126(33):10226-7.
[63] Coleman JN, Cadek M, Blake R, Nicolosi V, Ryan KP, Belton C, et al. High Performance Nanotube-Reinforced Plastics: Understanding the Mechanism of Strength Increase. Advanced Functional Materials. 2004;14(8):791-8.

[64]Tasis D, Tagmatarchis N, Georgakilas V, Prato M. Soluble Carbon Nanotubes. Chemistry - A European Journal. 2003;9(17):4000-8.
[65]Andreas H. Functionalization of Single-Walled Carbon Nanotubes. Angewandte Chemie International Edition. 2002;41(11):1853-9.
[66] Barber AH, Cohen SR, Wagner HD. Measurement of carbon nanotube--polymer interfacial strength. Applied Physics Letters. 2003;82(23):4140-2.
[67] Moniruzzaman M, Winey KI. Polymer Nanocomposites Containing Carbon Nanotubes. Macromolecules. 2006;39(16):5194-205.
[68]Liu L, Etika KC, Liao K-S, Hess LA, Bergbreiter DE, Grunlan JC. Comparison of Covalently andNoncovalently Functionalized Carbon Nanotubes in Epoxy. Macromolecular Rapid Communications.2009;30(8):627-32.
[69] Tasis D, Tagmatarchis N, Bianco A, Prato M. Chemistry of carbon nanotubes. Chemical Reviews.2006;106(3):1105-36.
[70] O'Connell MJ, Boul P, Ericson LM, Huffman C, Wang Y, Haroz E, et al. Reversible water-solubilization of single-walled carbon nanotubes by polymer wrapping. Chemical Physics Letters. 2001;342(3-4):265-71.
[71] Steuerman DW, Star A, Narizzano R, Choi H, Ries RS, Nicolini C, et al. Interactions between Conjugated Polymers and Single-Walled Carbon Nanotubes. The Journal of Physical Chemistry B. 2002;106(12):3124-30.
[72] Star A, Stoddart JF, Steuerman D, Diehl M, Boukai A, Wong EW, et al. Preparation and properties of polymer- wrapped single-walled carbon nanotubes. Angewandte Chemie - International Edition. 2001;40(9):1721-5.
[73] Zorbas V, Ortiz-Acevedo A, Dalton AB, Yoshida MM, Dieckmann GR, Draper RK, et al. Preparation and characterization of individual peptide-wrapped single-walled carbon nanotubes. Journal of the American Chemical Society. 2004;126(23):7222-7.
[74] Shim M, Javey A, Shi Kam NW, Dai H. Polymer Functionalization for Air-Stable n-Type Carbon Nanotube Field-Effect Transistors. Journal of the American Chemical Society. 2001;123(46):11512-3.
[75] Pan B, Xing B. Adsorption Mechanisms of Organic Chemicals on Carbon Nanotubes. Environmental Science & Technology. 2008;42(24):9005-13.
[76] Doe C, Choi SM, Kline SR, Jang HS, Kim TH. Charged Rod-Like Nanoparticles Assisting Single-Walled Carbon Nanotube Dispersion in Water. Advanced Functional Materials. 2008;18(18):2685-91.
[77] Vigolo B, Penicaud A, Coulon C, Sauder C, Pailler R, Journet C, et al. Macroscopic Fibers and Ribbons of Oriented Carbon Nanotubes. Science. 2000 November 17, 2000;290(5495):1331-4.
[78] Star A, Stoddart JF, Steuerman D, Diehl M, Boukai A, Wong EW, et al. Preparation and Properties of Polymer-Wrapped Single-Walled Carbon Nanotubes13. Angewandte Chemie International Edition. 2001;40(9):1721-5. 68.
[79]Yuan WZ, Mao Y, Zhao H, Sun JZ, Xu HP, Jin JK, et al. Electronic Interactions and Polymer Effect in the Functionalization and Solvation of Carbon Nanotubes by Pyrene- and Ferrocene-Containing Poly(1-alkyne)s. Macromolecules. 2008;41(3):701-7.
[80] Morishita T, Matsushita M, Katagiri Y, Fukumori K. Noncovalent functionalization of carbon nanotubes with maleimide polymers applicable to high-melting polymer-based composites. Carbon. 2010;48(8):2308-16.
[81] Dieckmann GR, Dalton AB, Johnson PA, Razal J, Chen J, Giordano GM, et al. Controlled Assembly of Carbon Nanotubes by Designed Amphiphilic Peptide Helices. Journal of the American Chemical Society. 2003;125(7):1770-7.

[82] Kim O-K, Je J, Baldwin JW, Kooi S, Pehrsson PE, Buckley LJ. Solubilization of Single-Wall Carbon Nanotubes by Supramolecular Encapsulation of Helical Amylose. Journal of the American Chemical Society. 2003;125(15):4426-7.
[83] Zheng M, Jagota A, Semke ED, Diner BA, McLean RS, Lustig SR, et al. DNA-assisted dispersion and separation of carbon nanotubes. Nature Materials. 2003;2(5):338-42.
[84] Zheng M, Jagota A, Strano MS, Santos AP, Barone P, Chou SG, et al. Structure-Based Carbon Nanotube Sorting by Sequence-Dependent DNA Assembly. Science. 2003 November 28, 2003;302(5650):1545-8.
[85] Lodge TP. Block Copolymers: Past Successes and Future Challenges. Macromolecular Chemistry and Physics. 2003;204(2):265-73.
[86] Zou J, Liu L, Chen H, Khondaker SI, McCullough RD, Huo Q, et al. Dispersion of Pristine Carbon Nanotubes Using Conjugated Block Copolymers. Advanced Materials. 2008;20(11):2055-60.
[87] Itzhak R, Raichman D, Shahar Z, Frey GL, Frey J, Yerushalmi-Rozen R. Tailoring Triblock Copolymers for Dispersion of Individual, Pristine, Single-Walled Carbon Nanotubes in Organic Solvents. The Journal of Physical Chemistry C. 2010;114(9):3748-53.
[88] Shvartzman-Cohen R, Levi-Kalisman Y, Nativ-Roth E, Yerushalmi-Rozen R. Generic Approach for Dispersing Single-Walled Carbon Nanotubes: The Strength of a Weak Interaction. Langmuir. 2004;20(15):6085-8.
[89]Sluzarenko N, Heurtefeu B, Maugey M, Zakri C, Poulin P, Lecommandoux S. Diblock copolymer stabilization of multi-wall carbon nanotubes in organic solvents and their use in composites. Carbon. 2006;44(15):3207-12.
[90] Cotiuga I, Picchioni F, Agarwal US, Wouters D, Loos J, Lemstra PJ. Block-Copolymer-Assisted Solubilization of Carbon Nanotubes and Exfoliation Monitoring Through Viscosity. Macromolecular Rapid Communications. 2006;27(13):1073-8.
[91] Lee JU, Huh J, Kim KH, Park C, Jo WH. Aqueous suspension of carbon nanotubes via non-covalent functionalization with oligothiophene-terminated poly(ethylene glycol). Carbon. 2007;45(5):1051-7.
[92] Shvartzman-Cohen R, Nativ-Roth E, Baskaran E, Levi-Kalisman Y, Szleifer I, Yerushalmi-Rozen R. Selective Dispersion of Single-Walled Carbon Nanotubes in the Presence of Polymers: the Role of Molecular and Colloidal Length Scales. Journal of the American Chemical Society. 2004;126(45):14850-7.
[93] Nativ-Roth E, Shvartzman-Cohen R, Bounioux C, Florent M, Zhang D, Szleifer I, et al. Physical Adsorption of Block Copolymers to SWNT and MWNT: A Nonwrapping Mechanism. Macromolecules. 2007;40(10):3676-85.
[94] Milton W. The Theory of Composites; 2002.
[95] McClory CC, S. J., McNally T. Polymer/Carbon Nanotube Composites. Australian Journal of Chemistry. 2009;62(8):762-85.
[96] Jin L, Bower C, Zhou O. Alignment of carbon nanotubes in a polymer matrix by mechanical stretching. Applied Physics Letters. 1998;73(9):1197-9.
[97] Bin Y, Kitanaka M, Zhu D, Matsuo M. Development of Highly Oriented Polyethylene Filled with Aligned Carbon Nanotubes by Gelation/Crystallization from Solutions. Macromolecules. 2003;36(16):6213-9.
[98] Wong M,Paramsothy M,Xu XJ,Ren Y,Li S,Liao K.Physical interactions at carbon nanotubepolymer interface. Polymer. 2003;44(25):7757-64.
[99] Ajayan PM, Stephan O, Colliex C, Trauth D. Aligned carbon nanotube arrays formed by cutting a polymer resin-nanotube composite. Science. 1994;265(5176):1212-4.
[100] Jin Z, Pramoda KP, Xu G, Goh SH. Dynamic mechanical behavior of melt-processed multi-walled carbon nanotube/poly(methyl methacrylate) composites. Chemical Physics Letters. 2001;337(1-3):43-7.

[101] Putz KW, Mitchell CA, Krishnamoorti R, Green PF. Elastic modulus of single-walled carbon nanotube/poly(methyl methacrylate) nanocomposites. Journal of Polymer Science, Part B: Polymer Physics. 2004;42(12):2286-93.
[102] Barraza HJ, Pompeo F, O'Rear EA, Resasco DE. SWNT-Filled Thermoplastic and Elastomeric Composites Prepared by Miniemulsion Polymerization. Nano Letters. 2002;2(8).
[103] Celzard A, McRae E, Deleuze C, Dufort M, Furdin G, Mar, et al. Critical concentration in percolating systems containing a high-aspect-ratio filler. Physical Review B. 1996;53(10):6209.
[104] Onsager L. THE EFFECTS OF SHAPE ON THE INTERACTION OF COLLOIDAL PARTICLES. Annals of the New York Academy of Sciences. 1949;51(4):627-59.
[105] Balberg I, Anderson CH, Alexander S, Wagner N. Excluded volume and its relation to the onset of percolation. Physical Review B. 1984;30(7):3933.
[106] Dai J, Wang Q, Li W, Wei Z, Xu G. Properties of well aligned SWNT modified poly (methyl methacrylate) nanocomposites. Materials Letters. 2007;61(1):27-9.
[107] Gorrasi G, Sarno M, Di Bartolomeo A, Sannino D, Ciambelli P, Vittoria V. Incorporation of carbon nanotubes into polyethylene by high energy ball milling: Morphology and physical properties. Journal of Polymer Science Part B: Polymer Physics. 2007;45(5):597-606.
[108] Dalmas F, Dendievel R, Chazeau L, Cavaillé J-Y, Gauthier C. Carbon nanotube-filled polymer composites. Numerical simulation of electrical conductivity in three-dimensional entangled fibrous networks. Acta Materialia. 2006;54(11):2923-31.
[109] Li C, Chou T-W. Continuum percolation of nanocomposites with fillers of arbitrary shapes. Applied Physics Letters. 2007;90(17):174108.
[110] GRUJICIC, 160, M., CAO, G., ROY, et al. A computational analysis of the percolation threshold and the electrical conductivity of carbon nanotubes filled polymeric materials. Heidelberg, GERMANY: Springer 2004.
[111] Stauffer D, Aharony, A. . Introduction to the Percolation theory. Taylor Francis et. London 1992.
[112] Clerc JP, Giraud G, Laugier JM, Luck JM. The electrical conductivity of binary disordered systems, percolation clusters, fractals and related models. Advances in Physics. 1990;39(3):191 - 309.
[113] Kilbride BE, Coleman JN, Fraysse J, Fournet P, Cadek M, Drury A, et al. Experimental observation of scaling laws for alternating current and direct current conductivity in polymer-carbon nanotube composite thin films. Journal of Applied Physics. 2002;92(7):4024-30.
[114] Du F, Fischer JE, Winey KI. Coagulation method for preparing single-walled carbon nanotube/poly(methyl methacrylate) composites and their modulus, electrical conductivity, and thermal stability. Journal of Polymer Science, Part B: Polymer Physics. 2003;41(24):3333-8.
[115] Gojny FH, Wichmann MHG, Fiedler B, Kinloch IA, Bauhofer W, Windle AH, et al. Evaluation and identification of electrical and thermal conduction mechanisms in carbon nanotube/epoxy composites. Polymer. 2006;47(6):2036-45.
[116] Gong X, Liu J, Baskaran S, Voise RD, Young JS. Surfactant-Assisted Processing of Carbon Nanotube/Polymer Composites. Chemistry of Materials. 2000;12(4):1049-52.
[117] Cui S, Canet R, Derre A, Couzi M, Delhaes P. Characterization of multiwall carbon nanotubes and influence of surfactant in the nanocomposite processing. Carbon. 2003;41(4):797-809.
[118] Stéphan C, Nguyen TP, De La Chapelle ML, Lefrant S, Journet C, Bernier P. Characterization of singlewalled carbon nanotubes-PMMA composites. Synthetic Metals. 2000;108(2):139-49.
[119] de la Chapelle ML, Stéphan C, Nguyen TP, Lefrant S, Journet C, Bernier P, et al. Raman

characterization of singlewalled carbon nanotubes and PMMA-nanotubes composites. Synthetic Metals. 1999;103(1-3):2510-2.
[120] Du F, Scogna RC, Zhou W, Brand S, Fischer JE, Winey KI. Nanotube networks in polymer nanocomposites: Rheology and electrical conductivity. Macromolecules. 2004;37(24):9048-55.
[121] Zhou Z, Wang S, Lu L, Zhang Y, Zhang Y. Preparation and rheological characterization of poly(methyl methacrylate)/functionalized multi-walled carbon nanotube composites. Composites Science and Technology. 2007;67(9):1861-9.
[122] Ferrara M, Neitzert HC, Sarno M, Gorrasi G, Sannino D, Vittoria V, et al. Influence of the electrical field applied during thermal cycling on the conductivity of LLDPE/CNT composites. Physica E: Low-Dimensional Systems and Nanostructures. 2007;37(1-2):66-71.
[123] Liang GD, Tjong SC. Electrical properties of low-density polyethylene/multiwalled carbon nanotube nanocomposites. Materials Chemistry and Physics. 2006;100(1):132-7.
[124] Valentino O, Sarno M, Rainone NG, Nobile MR, Ciambelli P, Neitzert HC, et al. Influence of the polymer structure and nanotube concentration on the conductivity and rheological properties of polyethylene/CNT composites. Physica E: Low-Dimensional Systems and Nanostructures. 2008;40(7):2440-5.
[125] McNally T, Pötschke P, Halley P, Murphy M, Martin D, Bell SEJ, et al. Polyethylene multiwalled carbon nanotube composites. Polymer. 2005;46(19 SPEC. ISS.):8222-32.
[126] Jin Z, Sun X, Xu G, Goh SH, Ji W. Nonlinear optical properties of some polymer/multi - walled carbon nanotube composites. Chemical Physics Letters. 2000;318(6):505-10.
[127] Goh HW, Goh SH, Xu GQ, Lee KY, Yang GY, Lee YW, et al. Optical Limiting Properties of DoubleC60-End-Capped Poly(ethylene oxide), Double-C60-End-Capped Poly (ethylene oxide)/Poly(ethylene oxide) Blend, and Double-C60-End-Capped Poly(ethylene oxide)/Multiwalled Carbon Nanotube Composite. The Journal of Physical Chemistry B. 2003; 107(25):6056-62.
[128] Abraham TN, Debdatta R, Siengchin S, Karger-Kocsis J. Rheological and thermal properties of poly(ethylene oxide)/multiwall carbon nanotube composites. Journal of Applied Polymer Science. 2008;110(4):2094-101.
[129] Chatterjee T, Yurekli K, Hadjiev VG, Krishnamoorti R. Single-Walled Carbon Nanotube Dispersions in Poly(ethylene oxide). Advanced Functional Materials. 2005;15(11): 1832-8.
[130] Musa I, Baxendale M, Amaratunga GAJ, Eccleston W. Properties of regioregular poly (3-octylthiophene)/multi-wall carbon nanotube composites. Synthetic Metals. 1999;102(1-3):1250-.
[131] Kymakis E, Alexandrou I, Amaratunga GAJ. High open-circuit voltage photovoltaic devices from carbon-nanotube-polymer composites. Journal of Applied Physics.2003;93(3): 1764-8.
[132] Kymakis E, Amaratunga GAJ. Single-wall carbon nanotube/conjugated polymer photovoltaic devices. Applied Physics Letters. 2002;80(1):112.
[133] Shaffer MSP, Windle AH. Fabrication and characterization of carbon nanotube/ poly (vinyl alcohol) composites. Advanced Materials. 1999;11(11):937-41.
[134] Ryan KP, Cadek M, Nicolosi V, Walker S, Ruether M, Fonseca A, et al. Multiwalled carbon nanotube nucleated crystallization and reinforcement in poly (vinyl alcohol) composites. Synthetic Metals. 2006;156(2-4):332-5.
[135] Cadek M, Coleman JN, Barron V, Hedicke K, Blau WJ. Morphological and mechanical properties of carbon-nanotube-reinforced semicrystalline and amorphous polymer composites. Applied Physics Letters. 2002;81(27):5123-5.
[136]Andrews R, Jacques D, Minot M, Rantell T. Fabrication of carbon multiwall nanotube /polymer composites by shear mixing. Macromolecular Materials and Engineering. 2002; 287(6):395-403.

MIX
Papier aus verantwortungsvollen Quellen
Paper from responsible sources
FSC® C105338

Printed by Books on Demand GmbH, Norderstedt / Germany